Space Science

Peter Pentland and
Pennie Stoyles

This edition first published in 2003 in the United States of America by Chelsea House Publishers, a subsidiary of Haights Cross Communications.

Chelsea House Publishers
1974 Sproul Road, Suite 400
Broomall, PA 19008-0914

The Chelsea House world wide web address is www.chelseahouse.com

Library of Congress Cataloging-in-Publication Data

Pentland, Peter.
Space science / by Peter Pentland and Pennie Stoyles.
v. cm. — (Science and scientists)

Includes index.
Contents: Have you ever wondered …? — Sight and light — Telescopes — Telescopes can find more than planets — How do you send things into space? — Rocket science — What is sent into space and why? — Probing the planets — What are satellites? — How does science help people survive in space? — What is it like to live in space? — Meet a NASA scientist — Is it possible to live on other planets? — Space science timeline.

ISBN 0-7910-7011-5
1. Space sciences—Juvenile literature. [1. Space sciences.] I. Stoyles, Pennie. II. Title.
QB500.22 .P46 2003
500.5—dc21

2002001284

First published in 2002 by
MACMILLAN EDUCATION AUSTRALIA PTY LTD
627 Chapel Street, South Yarra, Australia, 3141

Edited by Sally Woollett
Text design by Nina Sanadze
Cover design by Nina Sanadze
Illustrations by Pat Kermode, Purple Rabbit Productions

Printed in China

Acknowledgements
Cover: NASA space shuttle being launched, courtesy of Photoessentials.

AAP/AP Photo/Mikhail Metzel, p. 17 (top); JPL/TSADO/Tom Stack & Associates/Auscape, p. 18; Australian Picture Library/Corbis, p. 16 (top); Australian Picture Library/J. Carnemolla, pp. 10–11 (bottom); Dr Vaughan Clift, pp. 26–27; Getty Images/Photodisc, pp. 5 (top right), 7, 9, 10 (left), 16 (bottom), 17 (bottom), 20; Mt Stromlo Observatory, The Research School of Astronomy & Astrophysics, Institute of Advanced Studies, The Australian National University, p. 8; NASA, pp. 5 (bottom right), 19, 21, 24, 25; NASA/Oxford Scientific Films/Auscape, pp. 4–5 (bottom); Photoessentials, pp. 12–13, 14–15; Photolibrary.com/Science Photo Library, p. 22.

While every care has been taken to trace and acknowledge copyright the publisher tenders their apologies for any accidental infringement where copyright has proved untraceable.

Contents

Glossary words

When a word is printed in bold you can look up its meaning in the Glossary on page 31.

Science terms

When a word appears like this **dissolved** you can find out more about it in the science term box located nearby.

Have you ever wondered...

...what a satellite is?

...how space rockets are launched?

...if people will ever move from Earth to live on another planet?

...how people survive in space?

Did you know that all the answers have something to do with science?

What is space science?

Space science is a set of knowledge and theories that people have gathered over the years about our planet, the **solar system** and the universe. Curiosity about the stars and planets has caused the development of many fields of science.

With only their eyes to guide them, early civilizations made up explanations for happenings in the sky. Some thought that the stars were holes in a sphere that surrounded the world. The ancient Egyptians thought the sky was the body of a goddess and Earth was the body of a god.

Astronomy was perhaps one of the first sciences to be developed. People created a way of telling the time before clocks and calendars were invented. The time of day could be measured by the position of the Sun in the sky. The time of year could be measured by looking at the position of the stars in the night sky.

Scientists

Scientists try to explain why things happen. For thousands of years astronomers have tried to find out why the stars and planets move the way they do through the skies. More recently, scientists have made rockets and spacecraft to take people into space and to the Moon. Space probes have been sent to the planets and beyond the solar system. Teams of scientists are working to find ways to colonize other planets or their moons. Others seek to discover intelligent life in the universe.

There are many types of scientists, and they all have different jobs to do.

- Astronomers observe the stars and planets.
- Astrophysicists find out how stars behave.
- Hundreds of different kinds of scientists and engineers work on space travel, exploration and satellites.

In this book you will:

- find out how scientists discover planets, distant stars and galaxies
- learn how and why scientists send satellites and people into space
- read about how people survive in the unfriendly environment of space
- meet someone who works in the space industry and find out how he got his job.

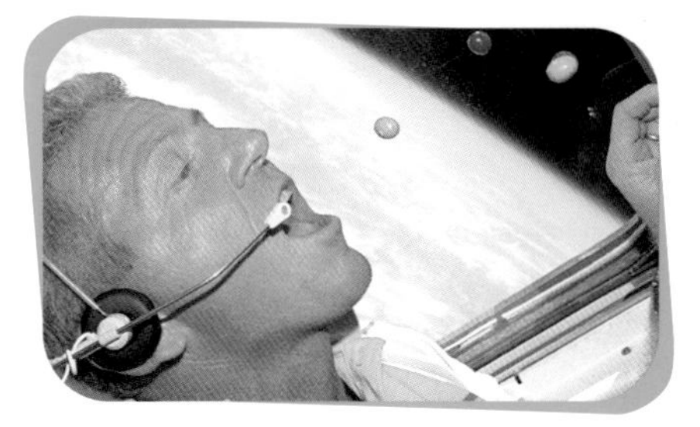

Sight and light

Have you ever wondered how your eyes work? What is light? Can you see light? What does light do?

Understanding sight and light is very important in helping to understand how discoveries were made in space science. You can see stars and planets that are huge distances away because they produce or **reflect** light.

Light is a type of **energy**. It is the only type of energy that human eyes are sensitive to. You can see things that produce light, reflect light or let light pass through them. You can see the flame of a candle because it gives off light. You can see a brick because light reflects off it. You can see through a glass bottle because light passes through it.

Science term

Energy is the ability of an object to do work. Energy cannot be created or destroyed, but it can be changed from one form to another.

How the eye works

- The cornea is a clear screen at the front of the eye that protects the eye and directs light at the lens.
- The pupil is a hole in the center of the iris that controls the brightness of the image and lets the light through to the lens.
- The lens can change shape to form a clear image of near or distant things.
- The retina is a screen at the back of the eye where the images form.
- Special cells in the retina called rods and cones change the images into messages. The rods respond to the brightness of the image and the cones respond to different colors.
- The messages are taken to the brain by the optic nerve. The brain processes the messages so you can make sense of what you see.

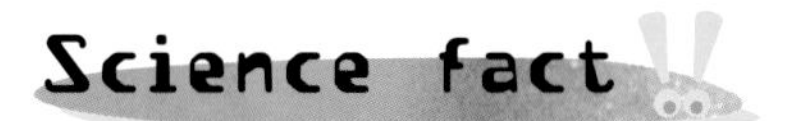

More than meets the eye

- Human eyes have 150 million rods and 10 million cones in each retina.
- Humans can tell the difference between about 10 million different colors.

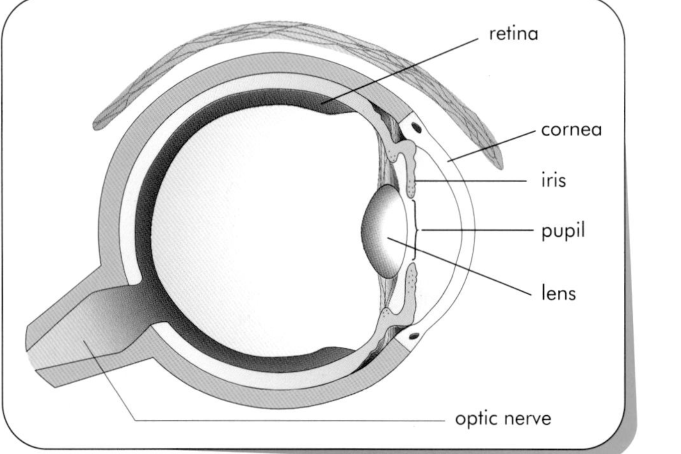

The human eye is made up of many parts.

What is light?

Light is a form of energy. Light can be changed into other forms of energy such as heat and electricity.

How fast is light?

Light is the fastest thing there is. It travels almost 200,000 miles (300,000 kilometers) in 1 second. Even traveling at this speed it takes 8 minutes for light to travel from the Sun to Earth. It takes light about 4 years to travel to us from the closest star (apart from the Sun).

Because light travels so quickly, scientists use a special measurement for distances in space, called a **light year**.

Science term

A light year is the distance that light travels in one year. This is a distance of about 6 trillion miles (9.5 trillion kilometers).

Light can be changed into electricity

Energy can be changed from one type to another. Spacecraft can use light from the Sun to produce electricity. If light hits a **solar panel**, some of its energy will be changed into electrical energy. Solar panels are used to give power to satellites and space stations.

The International Space Station gets some of its power from solar panels.

Telescopes

Long before the telescope was invented, people knew that other planets existed. They were also able to work out that the Sun and not Earth was the center of what is now called the solar system. How were they able to do this? How do telescopes work and how did their invention advance astronomy?

Scientists without telescopes

Before the invention of telescopes, scientists called astronomers made careful observations of the sky and kept very accurate records. They recorded the way that planets changed their positions relative to the stars.

The telescope at the Mount Stromlo Observatory in Canberra, Australia, looks different than early telescopes, but it works in the same way.

Nicolaus Copernicus (1473–1543) was a Polish astronomer. He lived at a time when it was generally believed that the stars and planets circled around Earth. Copernicus said that the motion of the planets through the sky could be explained by considering that Earth and other planets circled the Sun. The Greek astronomer Aristarchus of the island of Samos had come up with the same idea as Copernicus 1800 years earlier, but he had been unable to prove it. Later, astronomers such as Tycho Brahe (1546–1601) and Johannes Kepler (1571–1630) were able to accurately record these observations and make laws of planetary motion.

The first telescopes

Hans Lippershey, a Dutch spectacle maker, invented the telescope in 1608. His invention enabled scientists like Galileo Galilei (1564–1642) to make discoveries that were impossible using the eyes alone.

Famous scientist

Galileo Galilei

Galileo Galilei was an Italian astronomer and **physicist**. He used a telescope to observe four moons in orbit around Jupiter, craters and mountain ranges on the Moon, the phases of Venus, and the Milky Way. What he saw with his telescopes convinced him that Copernicus had been right. He also used it to study the surface of the Sun, studying such features as sunspots.

Galileo became blind four years before he died. His blindness was not because of his study of the Sun, which he did nearly 25 years before he went blind.

Types of telescopes

A telescope collects light and makes an image. An eyepiece is used to view and further magnify the image. There are two types of telescope, **refracting** and **reflecting**.

Refraction is the bending of light as it goes from one thing into another, for example from air into water.

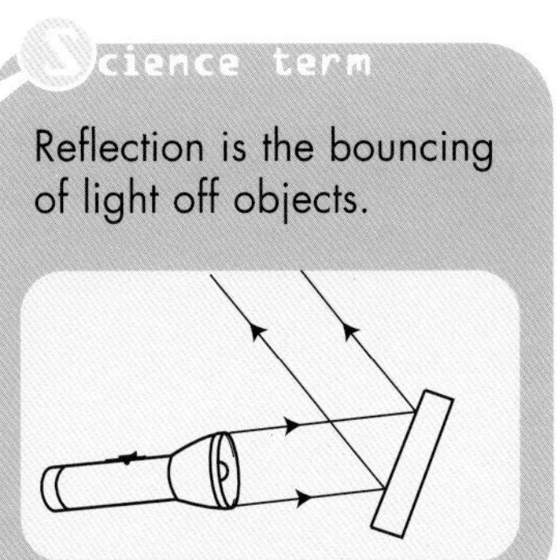

Reflection is the bouncing of light off objects.

Refracting telescopes

The first telescopes were called refracting telescopes because they used a **lens** to gather the light from the object being observed.

Reflecting telescopes

Reflecting telescopes use a **concave** mirror to gather the light. The light is reflected from the mirror and passes through the eyepiece.

Reflecting telescopes have some advantages over refracting telescopes.

- They can collect much more light than refracting telescopes. This makes them better for seeing very faint stars.
- It is much easier to make a big concave mirror than a lens of the same size.

The Hubble space telescope orbits around Earth.

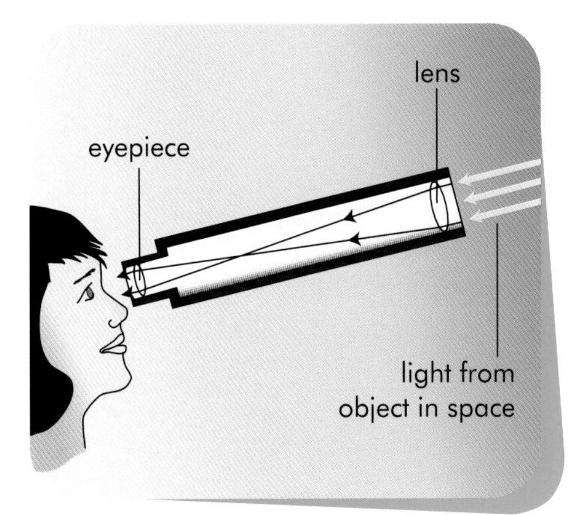

A refracting telescope uses lenses to refract light.

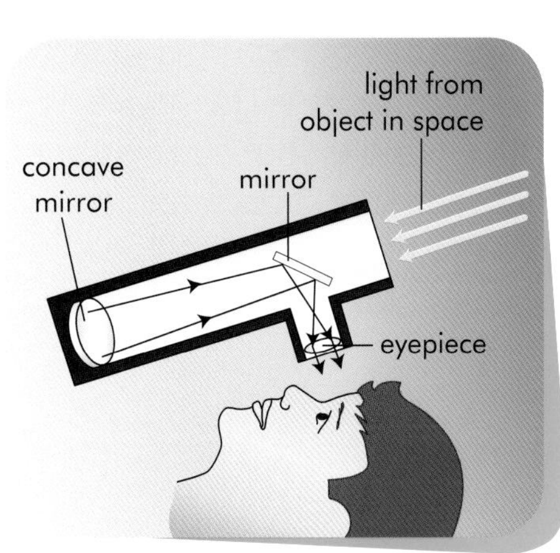

A reflecting telescope uses mirrors to reflect light.

Telescopes can find more than planets

What kinds of things have scientists been able to discover about space using refractive and reflective telescopes? What other things have been discovered in space? What instruments did scientists use to gather this information?

The Crab Nebula (a) and the Andromeda Galaxy (b) can both be seen using refractive and reflective telescopes.

Discoveries using ordinary telescopes

With the help of refractive and reflective telescopes scientists were able to discover the members of our solar system. These are the planets and their moons, the asteroids, the comets and meteoroids. They were even able to observe stars, galaxies and dust clouds called nebulae outside our solar system.

Discoveries using other types of telescopes

Scientists called astrophysicists have discovered a lot about stars. They have found out what stars are made of, how they produce their energy, how they are born, and how they die. They have also been able to gather evidence about many other objects beyond the solar system. These objects include pulsars, quasars and black holes. They have even been able to find evidence to support the theory that all space and time began about 15 billion years ago with an unimaginably huge explosion called the Big Bang.

What do stars produce?

Stars do not only produce light. Scientists have discovered stars produce other types of energy such as radio waves, microwaves, ultraviolet rays and x rays.

Radio waves from space!

A scientist named Karl Jansky (1905–1950) discovered radio waves from space in 1932. His discovery happened while he was investigating radiation that was interfering with radio communication. He found that the radio noises were coming from the center of our galaxy, the Milky Way.

Scientists were excited by Jansky's discovery and soon developed a telescope to investigate radio waves from outer space. This telescope is called a radio telescope. It gathers radio waves from space. It has a big dish that collects the waves and concentrates them onto an antenna. The antenna changes the radio waves into an electrical signal. This signal is made stronger by an amplifier, and a computer analyzes the signal.

Radio telescopes are also used for tracking space probes and collecting their radio signals.

X-ray telescopes

Stars give off some of their energy in the form of x rays. X-ray telescopes are used to find and measure x rays from distant sources such as stars. X-ray telescopes have been sent into orbit around Earth. The x rays produce an image that scientists can use to learn about distant objects in space.

This radio telescope is located at Parkes in New South Wales, Australia.

How do you send things into space?

Scientists have found out a lot about space by using different types of telescopes. To find out more they have sent instruments and people into space to take measurements. How did they do this? What holds everything onto Earth? What do you need to do to get an object into space? Once an object gets into space why does it stay there?

Science term

A force is a push or pull. It can change the movement of an object.

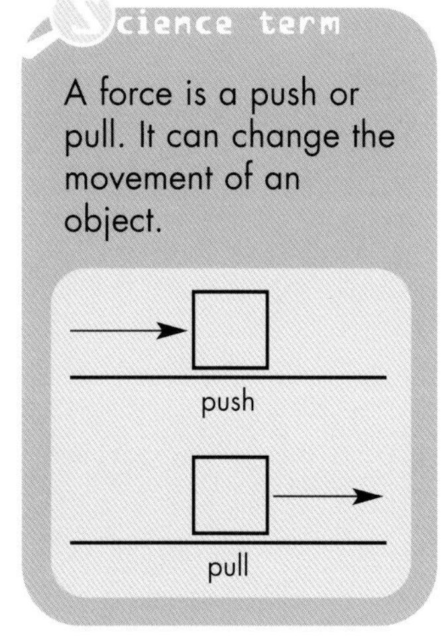

Force

Scientists have discovered that huge **forces** are needed to send spacecraft into space. These forces are needed to overcome the gravity that pulls everything toward Earth.

Gravity

Gravity is a force. It pulls masses together and even holds the atmosphere to Earth. Gravity keeps the Moon in orbit around Earth and stops it from flying off into space. It keeps Earth and the other planets orbiting around the Sun, and it holds galaxies together.

People usually refer to how heavy an object is by its weight. An object's weight takes into account the force of gravity pulling it toward the center of Earth. If you want to lose weight, all you have to do is move to a planet or moon that supplies a weaker force of gravity than Earth!

Try this

Can you calculate what your weight would be if you landed on the Moon? You can do this by measuring your 'Earth weight' using a bathroom scale and then dividing by six. You divide by six because the strength of gravity on the Moon is about one-sixth that of Earth.

You have less weight on the Moon.

Energy

You cannot make energy, but you can change it from one type to another. When you have a lot of energy you can run around and do lots of things. This type of energy is called kinetic energy. Potential energy is another type of energy. It is stored energy that an object has due to its shape or its position.

Scientists working on the first spacecraft realized that they needed a machine that could provide enough energy to push a spacecraft into space. The machine they invented was the rocket.

A rocket needs fuel so that it has the energy to push spacecraft. When the fuel burns, some of its energy is changed into the kinetic energy of the rocket and some is changed into the potential energy of the rocket as it moves farther above the surface of the planet.

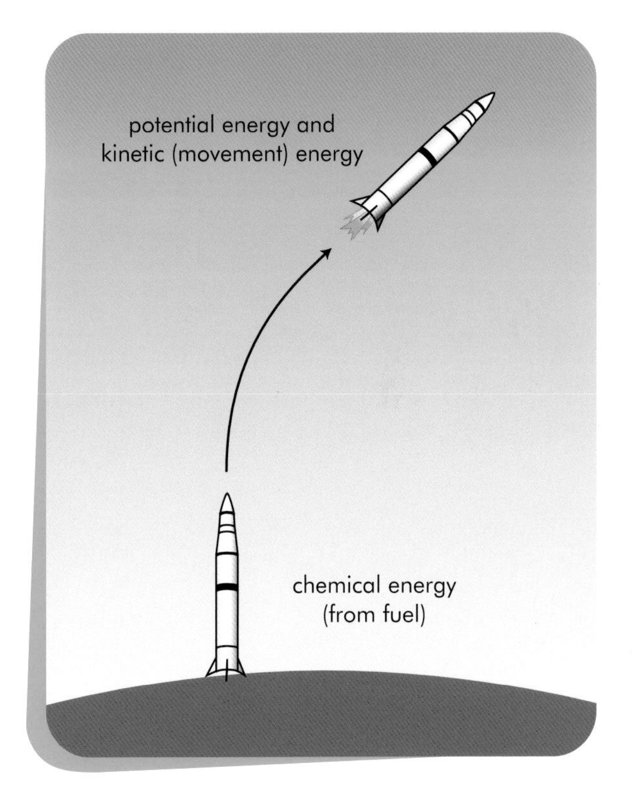

When a rocket is launched, some of the fuel's energy is changed into the kinetic energy and potential energy of the rocket.

A Saturn V rocket gives the force to lift a spacecraft into orbit. It gets its energy by burning fuel.

Rocket science

Scientists use rockets to launch spacecraft into space. Have you ever wondered how rockets work in space when there is no air to help the rocket fuel burn?

How rockets work

Rockets work by burning fuel to create large volumes of hot gases. The gases are pushed from the rocket in one direction and this causes the rocket to be pushed in the other direction.

Rockets do not need air to work. There is no air in space. Rockets use a chemical called an oxidizer to make the fuel burn.

The gases from a rocket are pushed in one direction and the rocket is pushed in the other.

Different types of rocket fuel

There are three general types of fuel:

- liquids such as gasoline that you burn in a car motor
- solids such as wood that you burn in a fireplace or a stove
- gases such as natural gas that you burn in a gas stove.

Rockets use fuels that are either solid or liquid.

The **NASA** space shuttle uses both solid and liquid fuels.

Solid fuel rockets

Solid fuel rockets do not have an engine. The fuel and oxidizer are ground up into a powder and then mixed together. When the powder burns it produces very hot gases that are pushed out of the nozzle. The fuel burns at a rate that gives the right amount of push on the rocket.

The NASA space shuttle uses two solid fuel rockets called boosters when it is launched. Each of the shuttle's solid rockets burns more than 11,000 pounds (5,000 kilograms) of fuel every second. They burn for two minutes. The solid rockets are separated from the shuttle using explosives when the fuel is used up. They parachute back into the ocean and are then recovered for use at another time. When the solid fuel boosters are used up, the shuttle is carried into orbit by its own liquid fuel rockets.

The fuel in a solid fuel rocket is made of a special powder.

Liquid fuel rockets

Liquid fuel rockets are more complicated than solid fuel rockets. They need:

- a special chamber where the fuel and oxidizer burn
- tanks to hold the fuel and oxidizer
- pumps to get the fuel and oxidizer into the chamber.

Liquid fuels are usually very cold. The NASA space shuttle uses liquid hydrogen as fuel and liquid oxygen as an oxidizer. These fuels are colder than –290 degrees Fahrenheit (–180 degrees Celsius).

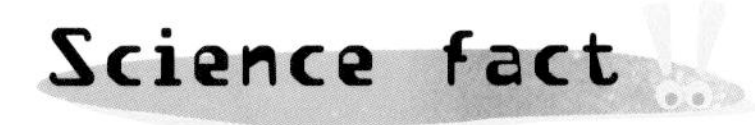

Freezing temperatures

One of the uses of negative (or 'minus') numbers is to describe temperatures that are colder than 0 degrees. Minus numbers are shown by putting a minus sign (–) in front of the number.

What is sent into space and why?

What sorts of things are launched into space? Why are they sent? Who was the first person into space? Who were the first people to walk on the Moon?

Laika was the first animal to travel into space.

People first walked on the Moon in 1969.

Venturing into space

Once scientists had developed the theory of rockets their next goal was to use the rockets to send people and instruments into space. The theory seemed simple enough. All that had to be done was to put a person or the instruments in a container on top of a rocket and set it off. The rocket provided enough force to lift itself and the spacecraft into space, a distance of at least 94 miles (150 kilometers).

Early space flights

In 1957 the **Soviet Union** launched *Sputnik 1*, the first vehicle into space. It was about the size and shape of a beach ball (a sphere shape) and carried a radio transmitter that sent a 'beep-beep' signal so that it could be tracked from the ground.

Later that year, the first animal traveled into space. The Russian dog Laika rode in the *Sputnik 2* satellite.

The first person to travel into space was the Soviet **cosmonaut** Yuri Gagarin in 1961. He rode in a sphere that was about 6 feet (2 meters) wide on top of a two-stage rocket called the *Vostok 1*. His flight lasted one hour and 48 minutes, and he reached a height of 204 miles (327 kilometers) above Earth.

The first Moon walk

One of the main reasons nations began to send people into space was to be the first to land people on the Moon. The first people to walk on the Moon were American astronauts Neil Armstrong and Edwin Aldrin. They landed the lunar lander *Eagle* on the Moon on July 20, 1969.

Space shuttles

The United States first launched a space shuttle in 1981. The Soviet Union followed with their shuttle in 1988. Space shuttles are reusable spacecraft. They are carried into space using liquid-fueled rocket engines. Rocket fuel is carried into space in a huge tank that is released after the fuel has been used. When the space shuttle has finished its work, it returns to Earth and can be used again. American shuttles also use two solid-fuel boosters to give extra force and energy to the shuttle.

The space rocket *Soyuz* was launched in Kazakhstan and took a crew to the International Space Station in 2000.

Space shuttles have many uses. They:

- carry satellites into orbit
- bring other satellites back to Earth
- allow repairs to be carried out on satellites
- carry crews to space stations
- provide a place for experiments

Space shuttles are slowed down by the atmosphere on reentry and use parachutes to slow them down further before they land.

Space stations

Space stations are large laboratories and living quarters that orbit Earth. They consist of smaller pieces that are built on Earth and joined together in space. Scientists, equipment and fresh supplies are carried to the space station in space shuttles.

One of the first experiments scientists carried out on a space station was to see what effect a long stay in weightless conditions had on the human body. They did this experiment to find out if people could survive the long trips needed to visit other planets.

The space station *Mir* went into space in 1986.

Science fact

Space visits

The record for the longest stay in space belongs to Russian doctor Valery Polyakov who returned to Earth on March 22, 1995, after 439 days aboard the Russian space station *Mir.*

The oldest person to fly in space was astronaut John Glenn, who at 77 years old, rode on the space shuttle *Discovery* in 1998. Glenn also was the first American to orbit Earth in 1962.

Probing the planets

As soon as scientists knew how to send a craft into space, they turned their attention to sending space probes to the Moon and to the other planets. What were they looking for? What instruments did they send to the other planets? How did they get the measurements and photographs back to Earth? How did they make electricity to run the space probes?

What are space probes looking for?

Unoccupied space probes carry scientific instruments to collect data about the planets they visit. The most interesting instruments for nonscientists are the cameras that send back spectacular images of the planets.

The instruments measure the planets' **magnetic fields**, the cosmic rays near the planets, asteroid dust and other information about the atmosphere and structure of the planets. Scientists back on Earth use the measurements to find out more about the solar system and how it was formed.

So far the only planets that space probes have landed on are Venus and Mars. The space probes that landed on Mars were looking for water and for living things.

This image of a volcano on Venus was created using data from the *Magellan* space probe, which orbited Venus for four years.

Making electricity

Space probes carry generators to make electricity. They need electricity to work the instruments and to send information back to Earth. **Radioactive** substances drive the generators. Electricity is also stored in batteries that can be recharged by the generators.

Sending information back to Earth

Each space probe carries a tiny radio transmitter. The data is changed into an electrical signal. A radio transmitter sends the signal back to Earth. A dish-shaped aerial directs the radio waves toward Earth.

Radio telescopes on Earth are pointed toward the probe. They gather the radio signals, which are incredibly weak. The signals are fed into amplifiers and then changed back into data or pictures by computers.

Photo stop

Some of the space probes that landed on Mars carried buggies that were controlled from Earth. This was a very slow process. It took 20 minutes for pictures to reach Earth and then another 20 minutes for the next command to travel back to the buggy on Mars.

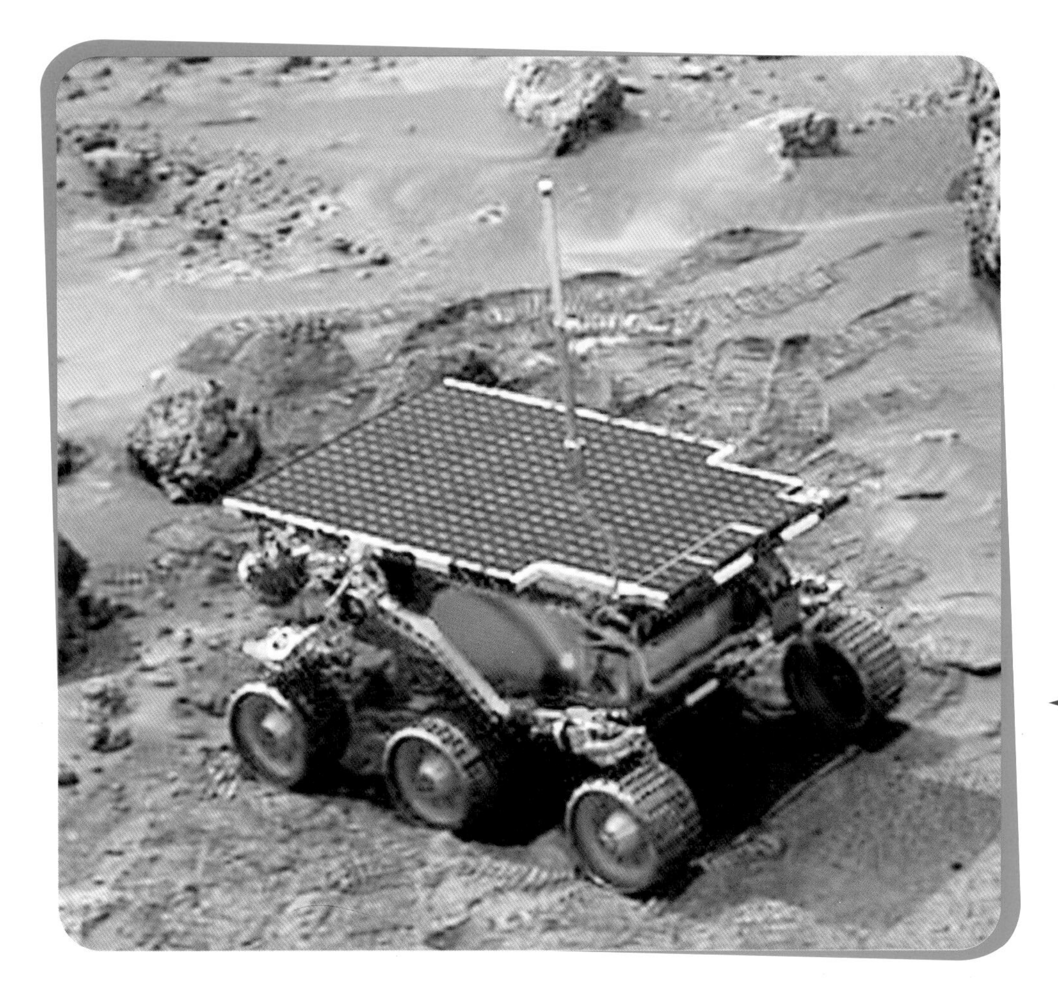

The Martian sojourner robot was controlled by signals sent from Earth.

Traveling beyond the solar system

The U.S. space probe *Pioneer 10* was launched in 1972. It traveled past Jupiter in 1973 and used the gravity of Jupiter to speed it up and send it out of the solar system. When it had been traveling for 25 years, it was twice as far from the Sun as Pluto. *Pioneer 10* carries a plaque that shows what people look like and visual instructions for finding Earth.

Pioneer 10 is heading for the **constellation** Taurus. It will pass a star there in about 2 million years.

What are satellites?

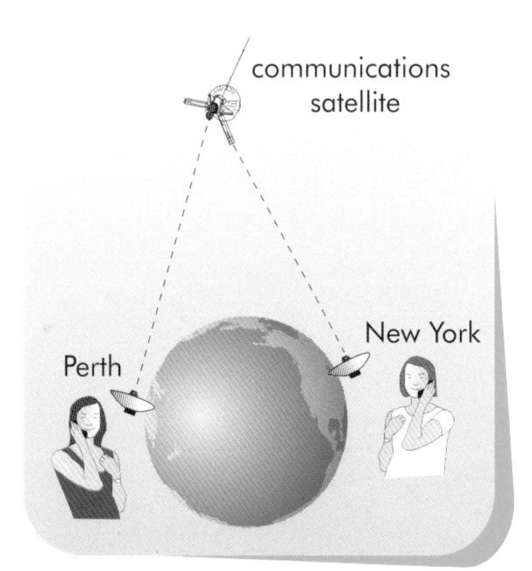

A phone call between New York and Perth, Australia, uses a satellite.

What are satellites? How did they get into space? What are they used for?

If you have ever seen a small dot of light traveling silently across the night sky it was probably an **artificial** satellite. You were able to see it because it reflected light from the Sun.

A satellite is any body that travels around a bigger body. The Moon is a natural satellite of Earth. The path that a satellite travels on as it goes around the bigger body is called its orbit.

How do satellites stay in orbit?

Scientists have put artificial satellites into orbit around Earth. Satellites have been carried into space on top of rockets or in space shuttles. They travel at heights of more than 125 miles (200 kilometers) above Earth's surface.

Satellites are held in orbit by the force of gravity that pulls them toward Earth. To stay in orbit, satellites must travel at great speeds. If a satellite is 125 miles (200 kilometers) above Earth's surface, it has a speed of nearly 5 miles (8 kilometers) per second. The farther above Earth's surface a satellite is, the slower it travels.

These astronauts are putting a satellite into orbit.

Different types of satellites

Satellites are used for many purposes. Here are some examples.

- Communications satellites help to send telephone calls around the world. They are usually in **geostationary** orbits. A message is sent from Earth to a satellite. The satellite sends the message to another satellite, which then sends it back to a receiving station on Earth.
- Earth-observing satellites such as *Landsat* carry instruments that record the temperatures and heights of land and sea masses.
- Weather satellites photograph clouds and measure temperatures.
- Spy satellites have powerful cameras to observe what is happening on the ground. The U.S. *Big Bird* spy satellite can photograph people. These satellites usually have low orbits. Some satellites listen to radio messages.
- Orbiting astronomical observatories carry different kinds of telescopes to observe our universe without the interference of the atmosphere. The Hubble space telescope is just one example.
- Global positioning satellites can give the position of a beacon to within a few feet. They can be used to locate sailors, hikers, aircraft and other travelers in isolated locations if something goes wrong.

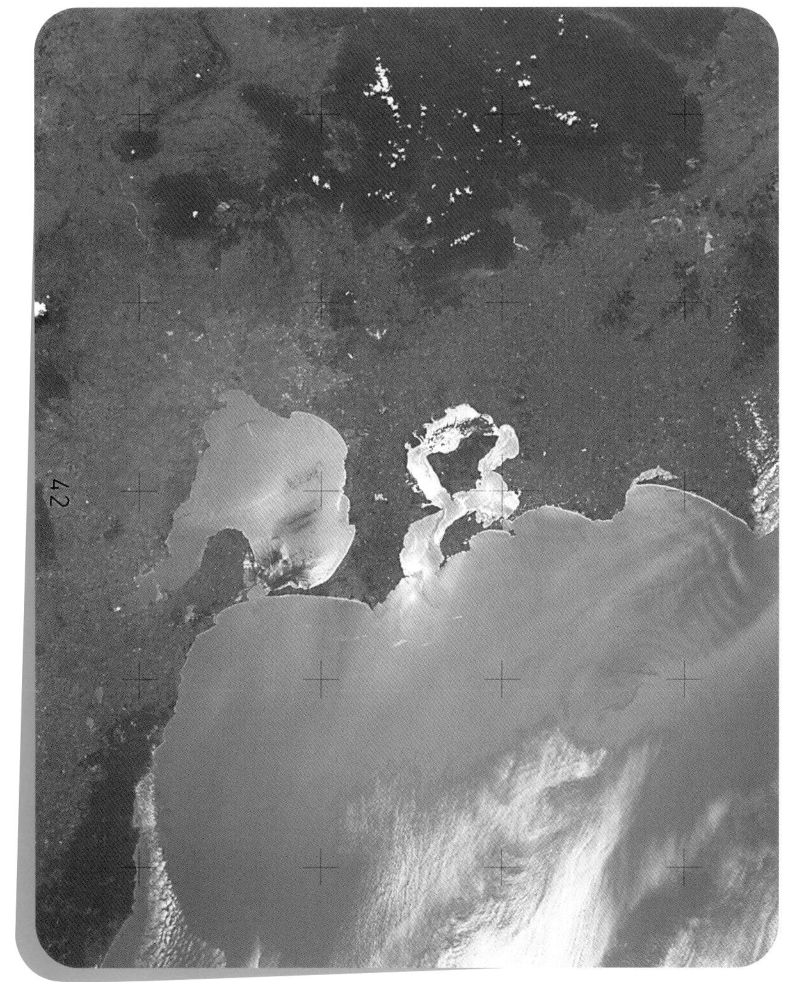

This photograph of Melbourne, Australia, was taken by a satellite.

- Space stations carry people and laboratories. The people carry out experiments in near-zero-gravity conditions. They investigate how living things behave in weightless conditions. They also investigate how physical processes and chemical reactions are affected by weightless conditions.
- Navigation satellites help ships and aircraft keep track of their positions.
- Space telescopes, such as the Hubble space telescope, take photographs and other measurements of objects and regions in space.

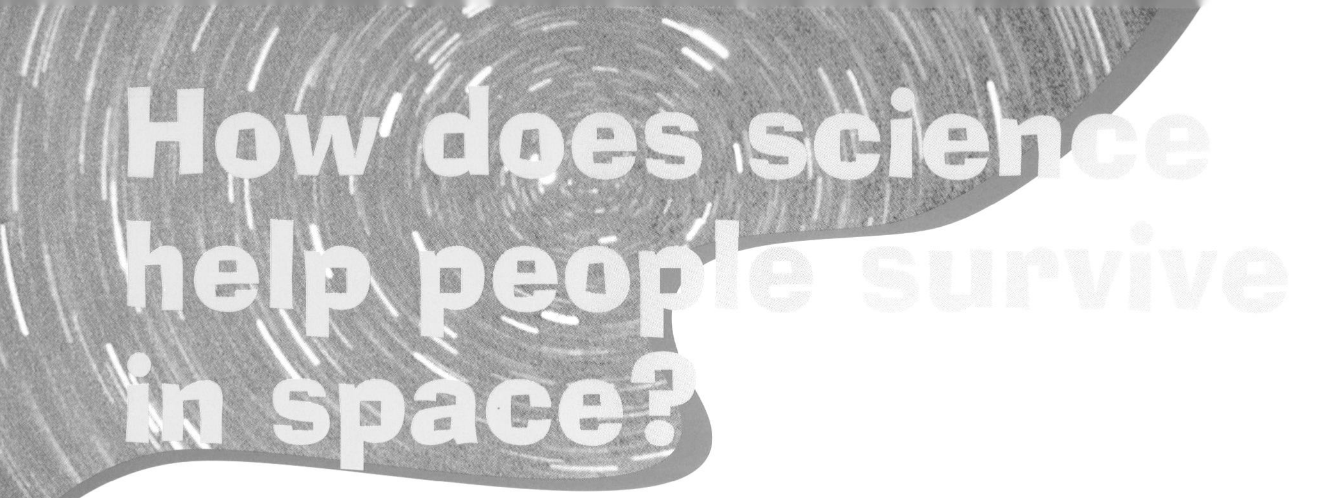

How does science help people survive in space?

What is it like in space? What do space travelers need to survive in space?

Space is not a friendly place to be. There is no air. It is not possible to breathe or to get oxygen without equipment. If space travelers ventured out into space without protection their blood would boil because there is no air pressure. Space is a place of extreme temperatures. When in orbit around Earth, the sides of spacecraft that face the Sun heat to as hot as nearly 250 degrees Fahrenheit (120 degrees Celsius). The sides that are in shadow are as low as –256 degrees Fahrenheit (–160 degrees Celsius).

Space scientists understand the conditions in space and the difficulties that visitors to space have. Their job is to carefully design and plan all the things that space travelers will need to be comfortable and safe.

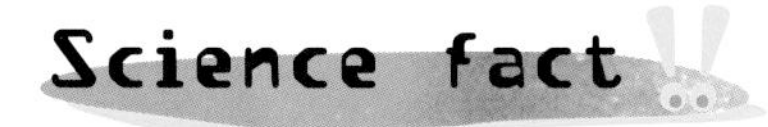

Are we there yet?

One of the main problems facing travelers to other planets in the future will be the psychological effect of being cooped up in a confined space for a long period of time. A one-way journey to Mars would take six months! Another problem will be how to deal with medical emergencies during the trip.

Space comfort

Space travelers need a sealed and pressurized environment that can support an Earth-like air supply. They need to have heaters and coolers to control the temperature inside the spacecraft. Temperatures are also controlled using insulation.

Food and drink

Spacecraft need to carry supplies of food and water for travelers to survive. Water is usually extracted from body wastes and used again.

Water has to be added to many space foods before they can be eaten.

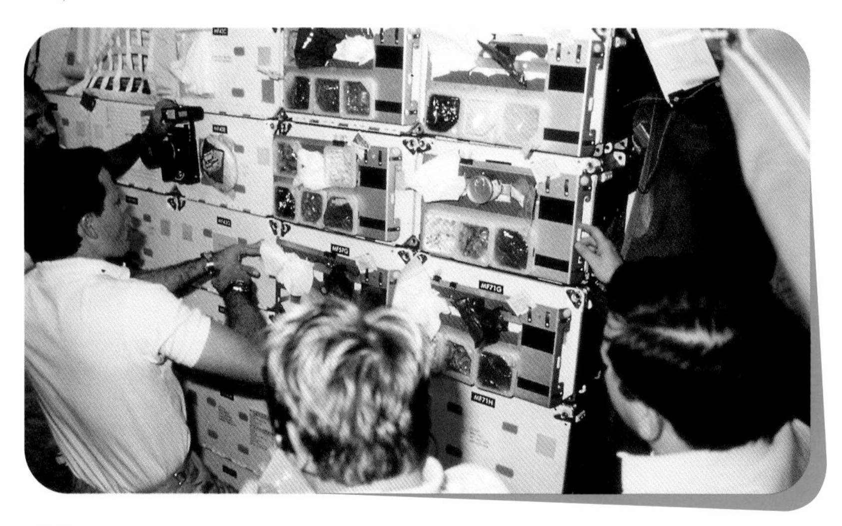

Breathing in space

Space travelers need oxygen to stay alive. They need to carry their own supply of oxygen or have a way of making oxygen. One way to make oxygen onboard a spacecraft is to pass electricity through water. This process is called electrolysis and produces both hydrogen and oxygen gases. The oxygen is stored in pressurized gas bottles.

Clean air

The air inside a spacecraft is cleaned using chemicals called scrubbers that naturally absorb water and other gases. They remove the **carbon dioxide** that people breathe out. People also breathe out **water vapor**, which has to be removed or it will quickly coat the surfaces inside the spacecraft. People produce other gases in their digestive systems that must be cleaned from the air.

Medical supplies and communication systems

Spacecraft need to carry medical supplies and a system to monitor the health of the people inside them. There needs to be a communication system between the craft and the command center back on Earth. If astronauts will be 'walking' outside the spacecraft, there also needs to be a communication system between the walker and the craft.

Electricity supply

All the systems need a supply of electricity to give them energy. Solar panels are placed on space stations to convert light energy into electrical energy. Fuel cells are also used to make electricity. They convert hydrogen and oxygen directly into electricity. Batteries store electrical energy for times when the other supplies are not working.

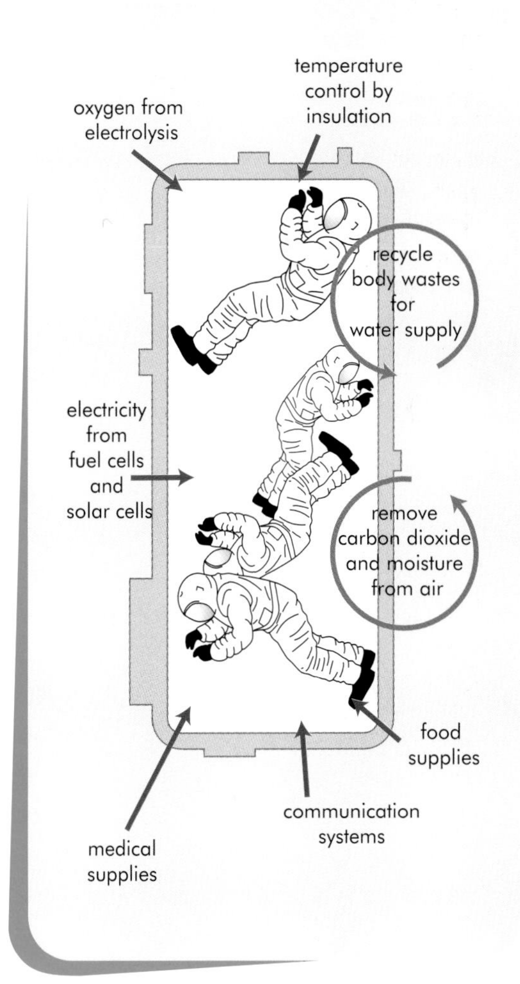

Many scientists worked together to design and build the life support systems on the International Space Station.

What is it like to live in space?

When space travelers orbit Earth in a spacecraft or space station, they have to get used to the experience of being weightless. What effects does weightlessness have on the human body? How does weightlessness affect the way space travelers do everyday things like eating and going to the toilet? What can they do to overcome any bad effects of living in space?

NASA astronauts train for weightlessness in a modified Boeing 707 jet, the KC-135. The jet is allowed to fall from a great height for 25 seconds at a time. The jet is called the 'vomit comet' because many people vomit when they first experience weightlessness.

What does weightlessness feel like?

Being weightless feels like you are constantly falling. In fact that is really what is happening to space travelers in orbit around Earth. When a spacecraft is in orbit, the spacecraft and everything inside it are 'falling' under the pull of gravity, so everything seems to be floating. The spacecraft is also traveling very fast so that it is able to keep a circular path and does not actually get closer to the planet below. If the spacecraft slowed down enough, it would indeed fall from space.

Effects of weightlessness on the body

Apart from making people feel sick, weightlessness has many other effects on the human body. Without the effects of gravity, fluids in the body move to the head, sometimes causing sinus pain. The semicircular canals in the middle ear use gravity to provide the sense of balance. They do not work in space and this sometimes causes **nausea**. Space travelers lose mass from their bones because their skeletons no longer have to support the weight of their bodies. They also lose muscle tone and mass because the muscles have less work to do. The heart is one muscle affected in this way.

Exercising, eating and sleeping are done differently in space.

Eating

Space travelers have to stop their bones getting weaker. Scientists called dieticians have worked out a diet that gives travelers food high in calcium, such as milk, cheese and yogurt. Food trays are held onto their laps with straps. Drinks come in soft plastic packs so that you cannot spill them.

Exercise

Exercise is important to stop people's muscles and bones becoming weaker. Scientific research has found that bones are made strong by doing weight-supporting exercises like walking and running. In weightless conditions people have to exercise on machines that reproduce gravity-like conditions. This is usually done using elastic straps that pull the body down onto a walking machine or treadmill.

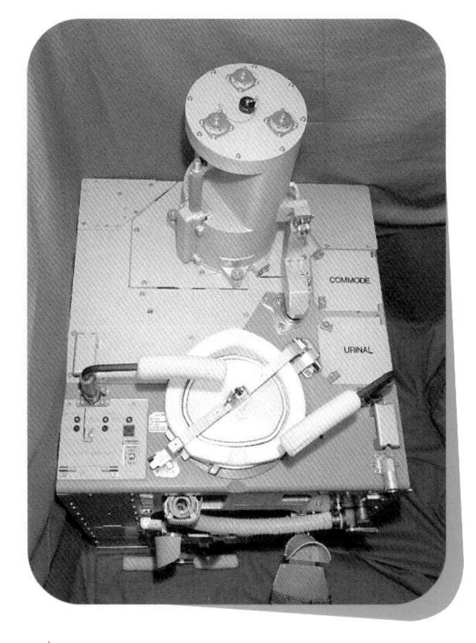

You have to anchor yourself on to the toilet in space.

Sleeping, washing and going to the toilet

Scientists such as engineers have to consider the everyday activities of space travelers when they are designing the inside of spacecraft.

You do not need a mattress when you are in orbit. People zip themselves into sleeping bags that are strapped to a wall so that they do not float around too much and bump into things.

People wash using wet wipes to prevent drops of water from floating around the spacecraft. In the toilet, solid and liquid body wastes are sucked into the disposal unit. People anchor themselves to the seat when they use the toilet so that they do not float away.

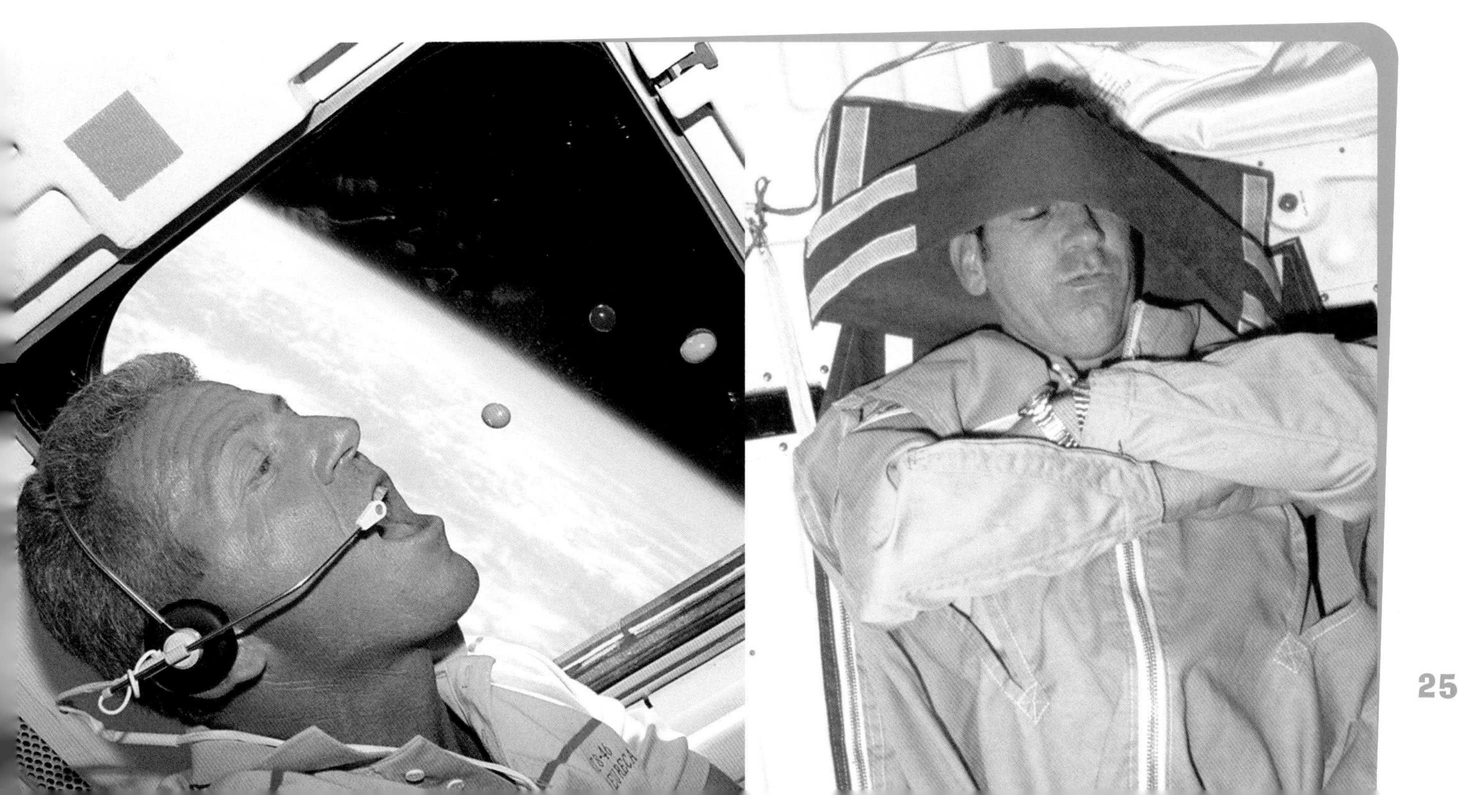

Meet a NASA scientist

Have you ever asked yourself what it would be like to work in the space industry? What sorts of jobs are there? How is it possible to get a job? What do you have to study in high school and at college?

Meet Vaughan Clift

Vaughan Clift has the answers to these questions. He is a scientist working for NASA. His job is in the field of medicine. He is responsible for monitoring the health of astronauts and keeping them alive in space.

Vaughan studied to be a doctor at university and did further training at two hospitals. At school he enjoyed studying physics and chemistry.

Vaughan's interest in science was stimulated by science fiction shows on television. His real inspiration to work in space science came when he was eight years old. He stood within a few feet of the Apollo 11 crew that had just returned from the first lunar landing.

Vaughan is a problem solver. In his job at NASA he alters the tools used in medicine so that they can work in the weightless conditions experienced in space travel. Vaughan gets to ride on the 'vomit comet' to test his equipment before it is sent into space.

Part of Vaughan's job is applying what he learns from his space work to monitoring and treating people back on Earth. He has been part of many discoveries about how the human body works. He says that human space flight is one way of exploring life on Earth and that space experiments teach us a lot about the human body.

Vaughan describes himself as a professional coward because he hates needles. He invented a way of measuring blood chemistry without putting needles into people. This invention is useful to NASA because some astronauts are also afraid of needles.

Vaughan says NASA is a great place to work because amazing events like shuttle launches are everyday topics. He finds that working with someone who later becomes an astronaut is a very exciting experience. Working at NASA gives people a chance to try to develop new ideas. It also provides new and special challenges.

Vaughan's advice is, 'Do not be afraid to dream. Many will tell you that you are dreaming if you think you can work in the space program or be an astronaut. It is not true. Be brave enough to admit to yourself that this is what you would like to do. Openly pursue your dream and you will succeed.

'The areas of space research are so broad that you can pursue almost any field of science. Do it well and develop a name for yourself in that field. Then the doors will open.'

Dr. Vaughan Clift with the instrument he invented to take blood samples from astronauts.

Is it possible to live on other planets?

Will people ever be able to live on other planets? What will a space colony look like? What does a space colony need to keep people alive?

Life on Mars

Scientists believe that Mars might be changed one day into a place where people can live. A process called **terraforming** will do this.

Science term

Terraforming is the changing of a planet's environment to produce Earth-like conditions.

Livable conditions

At the moment Mars is not a good place for people to live. This is because:

- it is too cold
- its atmosphere is too thin
- there is no liquid water
- its surface is exposed to harmful rays from space.

To make Mars suitable for people, these problems will have to be fixed. There needs to be more oxygen and nitrogen in the atmosphere so that it is Earth-like.

Spreading dark soil on the polar caps of Mars will start the process of changing its environment.

Warmth leads to an atmosphere

Scientists believe that Mars once had a much thicker atmosphere that was mainly made of carbon dioxide. They believe that most of this carbon dioxide is still present in solid form on the surface of the planet. If they can warm the planet a little it will release some carbon dioxide into the atmosphere. This will trap even more heat because the carbon dioxide will let light from the Sun pass through it, but it will not let heat get back out.

This process will make Mars warmer and release more carbon dioxide into the atmosphere, which will trap more heat. The atmosphere will be built up and the planet will become warmer.

Scientists believe that spreading Mars' dark-colored soil onto the polar icecaps (like the north and south poles on Earth) would start the warming process. Dark colors are good at absorbing light from the Sun.

The future for life in the solar system

When will ordinary people be able to take rides into space? Will there ever be holiday resorts on the Moon or on planets other than Earth? Could you ever have a job mining the asteroids? Will Earth become so overcrowded with people that they will have to move to other planets?

Making a space colony on a planet or moon would be a slow process. There would be lots of things for scientists to think about.

- A space station-like structure would be built first. It would get bigger as more parts were brought in. It would be safer to build most of the colony underground. This would protect the colony from meteors. Underground chambers could be sealed easily to keep the air in.
- Solar panels or fuel cells could provide electricity. Fuel cells are good to use because they also make water.
- Water and oxygen are essential for survival. They could be produced from chemicals that can be found in Mars's rocks and soil. Oxygen and hydrogen are among the most plentiful elements in the universe.
- Surface dwellings would be contained in a tough, see-through dome. This would allow plants to grow. Plants are important because they remove carbon dioxide from the air and release the oxygen people need to breathe. They also produce food and wood.

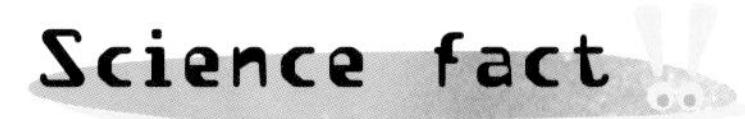

Cities in space

A team of NASA scientists is currently working on a project to build colonies on the Moon and Mars.

A space colony of the future might look something like this.

Space science timeline

This timeline shows some important space science events. See if you can imagine some of the things that might happen in space science in the future.

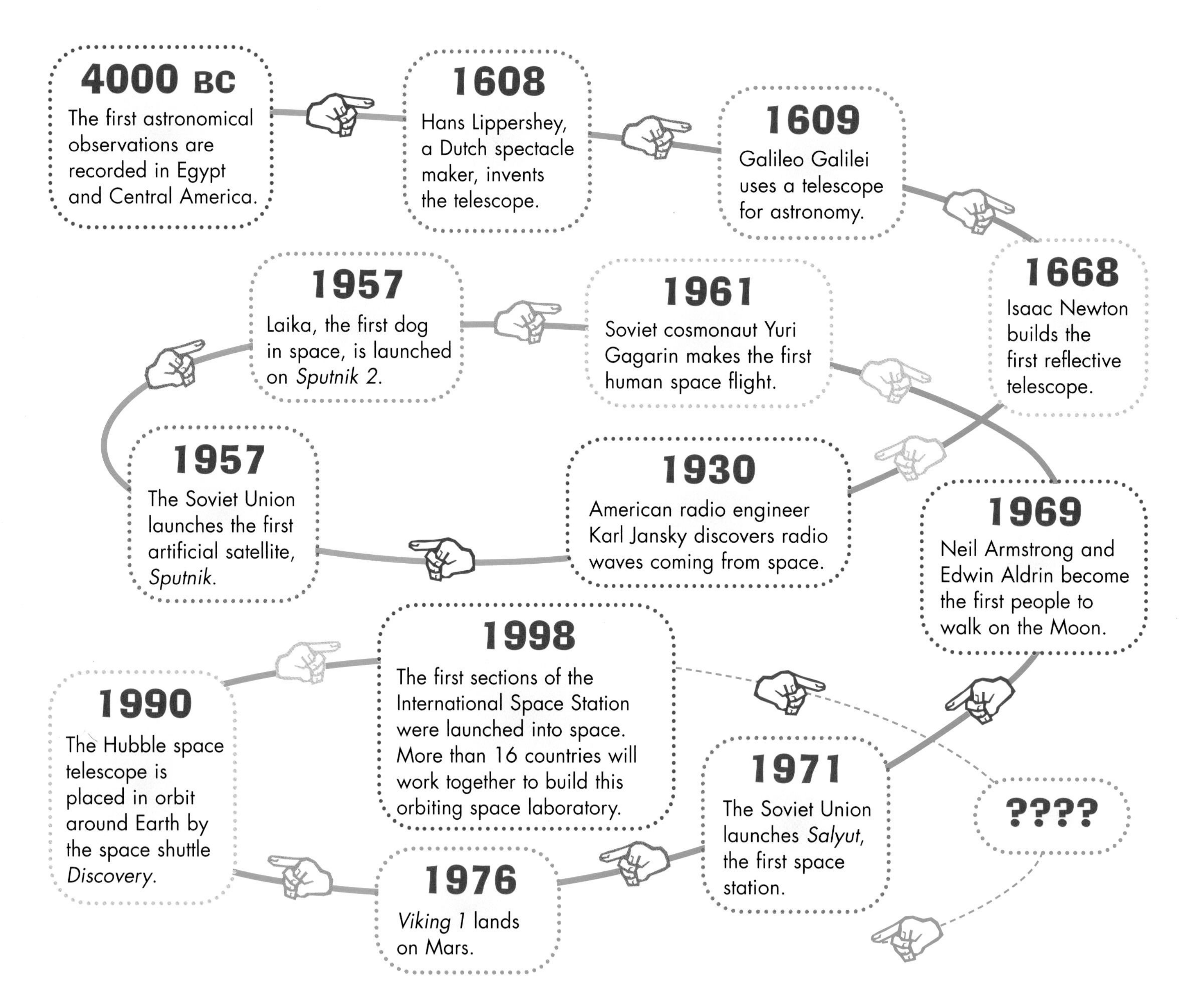

What are scientists working on now?

- Work has begun on the International Space Station.
- Scientists plan to launch a three-year NASA mission sending astronauts to Mars in about 2020.

Glossary

artificial not occurring naturally; made by people

carbon dioxide a gas that is produced by animals and taken in by plants

concave shaped like the inside of a ball

constellation a group of stars in a particular section of the sky

cosmonaut an astronaut from the Soviet Union

energy the ability of an object to do work. Energy cannot be created or destroyed, but it can be changed from one form to another

forces pushes or pulls; they change the movement of an object

geostationary in orbit around Earth, always directly above one place on Earth's equator

lens a specially shaped piece of glass that forms an image

magnetic fields regions where magnets will experience a force

NASA National Aeronautics and Space Administration, the American organization responsible for space exploration

nausea a desire to vomit

physicist a scientist who studies the way things behave and the way energy interacts with matter

radioactive giving off charged particles or high-energy rays

reflect to bounce light back from a surface

solar panel a collection of solar cells connected together to produce a desired amount of electricity

solar system the Sun and its family of planets, their natural satellites, comets, asteroids, and meteoroids

Soviet Union a union of states that no longer exists. Countries that were part of the Soviet Union include Russia and the Ukraine

water vapor the gaseous state of water

Index